XTREME AIRCRAFT

AEROBATIC AIRCRAFT

A&D Xtreme
BOLD HI-LO NONFICTION
An imprint of Abdo Publishing
abdobooks.com

S.L. HAMILTON

TAKE IT TO THE XTREME!

GET READY FOR AN XTREME ADVENTURE!
THE PAGES OF THIS BOOK WILL TAKE YOU INTO
THE THRILLING WORLD OF AEROBATIC AIRCRAFT.
WHEN YOU HAVE FINISHED READING THIS BOOK, TAKE THE
XTREME CHALLENGE ON PAGE 45 ABOUT WHAT YOU'VE LEARNED!

ABDOBOOKS.COM

Published by Abdo Publishing, a division of ABDO, PO Box 398166, Minneapolis, Minnesota 55439.

Printed in the United States of America, North Mankato, MN.

092021

012022

THIS BOOK CONTAINS RECYCLED MATERIALS

Editor: John Hamilton

Graphic Design: Sue Hamilton

Cover Design: Laura Graphenteen

Cover Photo: Shutterstock

Interior Photos & Illustrations: Aircraft Owners & Pilots Association-pg 18; Alamy-pgs 14-15; Extra Aircraft-pg 19; Game Composites-pgs 32 & 33; Granger-pgs 8-9; Pilots of America-pg 16; Red Bull Content Pool-pgs 20-21, 24-25, 26-27, 38-29, 42 & 43; San Diego Air & Space Museum-pgs 10 & 11; Shutterstock-pg 1, 4-5, 12, 13, 17, 22-23, 27 (inset), 30-31, 40 & 41; Smithsonian Institution-pgs 6 & 7; US Air Force-pgs 34-35; US Air National Guard-pgs 28 & 29; US Navy-pgs 36-37 & 44.

LIBRARY OF CONGRESS CONTROL NUMBER: 2021943583

PUBLISHER'S CATALOGING-IN-PUBLICATION DATA

Names: Hamilton, S.L., author.

Title: Aerobatic aircraft / by S.L. Hamilton

Description: Minneapolis, Minnesota : Abdo Publishing, 2022 | Series: Xtreme aircraft | Includes online resources and index.

Identifiers: ISBN 9781532197321 (lib. bdg.) | ISBN 9781098219529 (ebook)

Subjects: LCSH: Aviation--Juvenile literature. | Aerobatics--Juvenile literature. | Aerobatic flying--Juvenile literature. | Stunt flying--Juvenile literature.

Classification: DDC 629.1333--dc23

TABLE OF CONTENTS

AEROBATIC AIRCRAFT 4
HISTORY 6
AEROBATIC MANEUVERS 12
AEROBATIC PLANE CATEGORIES 16
TOP SINGLE-SEATERS 18
TOP DOUBLE-SEATERS 26
JETS & FORMATION FLYING 34
AEROBATIC HELICOPTERS 38
AEROBATIC RC PLANES 40
BECOMING AN AEROBATIC PILOT 42
THE FUTURE 44
XTREME CHALLENGE 45
GLOSSARY 46
ONLINE RESOURCES 47
INDEX 48

CHAPTER 1

AEROBATIC AIRCRAFT

Aerobatic aircraft are often flown to entertain. Flying upside down, close to the ground, brings many oohs and aahs from amazed crowds.

Aerobatic aircraft are planes and helicopters designed to fly tricky, often dangerous, stunts in the air. The term refers to aerial acrobatics. Pilots and their planes are often seen in air shows or competitions.

XTREME FACT

Aerobatic pilots use smoke to help crowds on the ground see the aircraft's stunts better.

CHAPTER 2

HISTORY

Wilbur and Orville Wright flew the first controlled flights of a powered airplane in 1903. They made the first banked turn in 1904, flexing the entire wing. When the aileron, a hinged section of the wing, was invented, turning was much easier and aerobatics became possible.

For the 1909 New York flying demonstration, Wilbur Wright attached a red canoe to his plane in case it went down in the water.

In 1909, Wilbur Wright was hired to perform a series of flights for a special New York celebration. He flew a modified plane around the Statue of Liberty and up to Grant's Tomb, about 20 miles (32 km) away. More than a million people watched. The event began the use of planes for entertainment.

World War I (1914-1918) increased the skills of pilots and the quality of planes. At first, the **fragile** planes were used for **reconnaissance** and dropping hand-held bombs. Later, as planes became more powerful and pilots more skilled, aerial **dogfights** took place in the sky. **Aces** faced off against each other, using steep climbs, dives, fast turns, and loops, to take out their airborne enemies.

XTREME FACT

World War I's wood-and-canvas planes were so unreliable that pilots joked they were "coffins with wings." In 1915, the average life expectancy of a pilot at the front lines was only 11 days.

After **World War I** ended, fewer military planes were needed. The Curtiss JN-4 "Jenny" **biplane** and the Standard J-1 were bought cheaply by the skilled pilots who still wanted to fly them. Air shows became popular.

Stunt performer Wally Timm hangs by his knees from a Jenny biplane.

Wing walkers Ivan Unger and Gladys Roy play a dangerous game of tennis in the air in 1925.

A group of aerobatic planes was called a flying circus. Individual pilots were **barnstormers**. People paid for rides on their planes. Pilots also performed daredevil shows, flying loops, spiral dives, upside down (inverted flight), or between trees or telegraph poles. Wing walkers entertained crowds with dances and acrobatic stunts.

CHAPTER 3

AEROBATIC MANEUVERS

Basic aerobatic moves are the building blocks of more complicated maneuvers. Aerobatic pilots first learn to do spins, rolls, loops, and inverted flying. All require master flying skills and a pilot and plane that can handle extreme **g-forces**.

A spin is a vertical twist in the air.

A loop requires the pilot to view the horizon and the ground at confusing angles.

Inverted flying requires the pilot to control the plane while upside down.

Rolls are horizontal twists in the air using the plane's ailerons.

Complicated aerobatic moves use a combination of basic maneuvers in unique and different ways. They often have colorful and descriptive names, such as a tailslide,

pinwheel, corkscrew, hammerhead, J-turn, lazy eight, and question mark. Some require the use of jet power to perform the moves, but all need precise timing and skill.

CHAPTER 4

AEROBATIC PLANE CATEGORIES

The Cessna Aerobat may be used for pilot training or for aerobatics.

Aerobatic planes may be divided into two categories. Planes built specifically for aerobatics are called specialist aerobatics. Those that may be used for other purposes, such as carrying passengers or touring, are called aerobatic capable aircraft.

A Sukhoi Su-26M is a specialist aerobatic plane with extreme power. Pilots have won several aerobatic championships flying it.

CHAPTER 5

TOP SINGLE-SEATERS

Germany's Extra 330SC is considered the best single-seat aerobatic aircraft. It is powerful and precise, with a steel frame and a carbon fiber skin. However, it is difficult to control and the cockpit is loud. Winners of many world aerobatic competitions fly this plane.

The cockpit of an Extra 330SC has a steel frame to add strength to the powerful plane.

The Extra 330SC's cockpit is behind the wings. This gives the pilot a better view of the ground to position the plane for precise aerobatic moves.

Zivko Aeronautics' Edge 540 is known for its speed and ability to handle high **g-forces**. The aerobatic plane is

The Edge 540 is flown by pilots in the Challenger Class of the Red Bull Air Race World Championships.

designed for fast turns and quick climbs. The Edge 540 has a top speed of 265 mph (426 kph), but is also very stable.

The Pitts Special S1 is a single-seat aerobatic **biplane**. It was designed by Curtis Pitts (1915-2005) and first flew in 1944. The US Aerobatic Team won the 1972 World Championships flying Pitts Specials. A Pitts Special **variant** called the Sunbird S1X holds the world record for the most inverted flat spins. Aerobatics pilot Spencer Suderman performed 98 spins in a row in 2016.

XTREME FACT

Several Pitts planes had pictures of skunks on them. The planes were nicknamed "Stinkers."

Pilots Paul Bonhomme and Steve Jones of the Red Bull Matador Aerobatic Team fly two XA41s in close formation, through a hangar 3 feet (1 m) off the ground at a speed of 185 mph (298 kph).

XtremeAir's Sbach 300, also known as the XA41, is an aerobatic plane with easy slow-speed handling and accurate high-speed maneuverability. The carbon-fiber plane has a top speed of 242 mph (389 kph).

CHAPTER 6

TOP DOUBLE-SEATERS

Some aerobatic aircraft hold two people. Russia's Sukhoi (pronounced soo-hoy) Su-29 is a double-seater with a top speed of 239 mph (385 kph). Originally designed as a training aircraft for the Russian military, it is nimble and powerful. It has a roomy cockpit with excellent visibility.

The Sukhoi Su-29's seats have a five-point harness, similar to what is worn by race car drivers.

The Sukhoi Su-29 is known for its ability to perform loops.

The Pitts S-2C's 3-blade aerobatic propeller is called the "Claw." It gives the plane better hang time, allowing the pilot extra seconds at the top of maneuvers.

Aviat Aircraft's Pitts S-2C is a two-seater **biplane** capable of rolling 300 degrees per second. With squared-off wings, sleek lines, and better pilot visibility, it is built for aerobatic competition and training.

The Yak-52 is a two-seater created by the famous Russian aircraft designer Aleksandir Yakovlev. Made of heavy-gauge aluminum, it's tough and powerful. The sturdy engine sends the plane climbing at a rate of up to 2,000-3,000 feet per minute (610-914 m per minute).

Yak-52s are nicknamed "Red Star" planes.

XTREME FACT

To be flown in the United States, the Yak-52's gauges and controls must have English translation labels from the original Russian Cyrillic writing.

XTREME FACT

Aerobatic aircraft sometimes lose their canopy, the clear bubble covering the pilot and cockpit. The GB1 has a split canopy, so if the canopy blows off, the pilot still has a small windshield for protection.

Game Composites' GB1 GameBird is one of the newest aerobatic aircrafts. The first plane was completed in November 2018. The GB1 is capable of incredible slow-speed maneuvers and has a roll rate of 450 degrees per second at 270 mph (435 kph). It's designed for fun!

The GB1 GameBird already had a second place win at the 2019 South African National Aerobatic Championships.

CHAPTER 7

JETS & FORMATION FLYING

Formation flying is when two or more aircraft fly with precise control. Military pilots and jets are often used. The US Air Force Thunderbirds fly in F-16 Fighting Falcons. The F-16's light weight and powerful engines allow for amazing formation maneuvers.

Two US Air Force Thunderbirds in F-16 Fighting Falcons fly in a mirror formation.

XTREME FACT

In early aviation, a group of aircraft flying together was called a "gaggle." Today, they are usually referred to as a fleet or squadron.

The leader in formation flying is usually the pilot with the most experience. All other pilots are known as wingmen.

The US Navy Blue Angels perform formation flying in F/A-18 Super Hornets. These strike fighters are powerful, nimble jets. Maneuvers are tight, well-practiced displays, with the jets often less than 3 feet (1 m) apart. During shows, pilots fly at speeds of 450-700 mph (724-1,127 kph). Their top speeds are just below the **sound barrier**, so no **sonic booms** are heard.

XTREME FACT

In formation flying, only the leader speaks on the radio to the other pilots. Wingmen communicate with nods, hand signals, or by using their aircraft.

CHAPTER 8

AEROBATIC HELICOPTERS

The MBB Bo-105 was the first aerobatic helicopter. Introduced in 1970, the Bo-105 featured a special **rotor** design that allowed it to climb quickly and perform rolls, loops, dives, and turnovers.

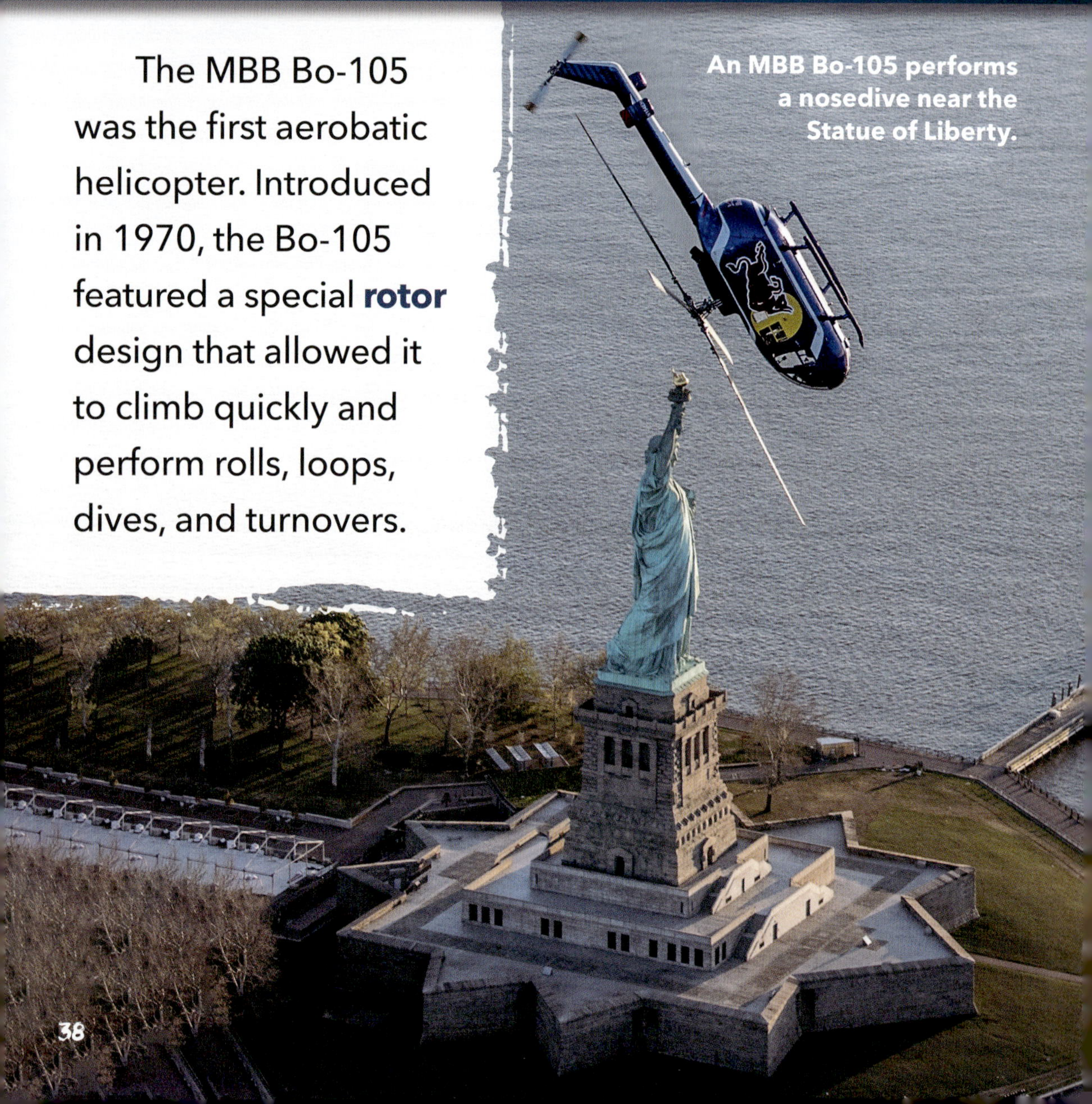

An MBB Bo-105 performs a nosedive near the Statue of Liberty.

A view of the New York City skyline from the cockpit of an inverted Red Bull aerobatic helicopter with pilot Aaron Fitzgerald.

XTREME FACT

There are only a few helicopters designed for aerobatics. Their original purpose was for military or police duties. Their light weight and maneuverability make them excellent rescue helicopters.

CHAPTER 9

AEROBATIC RC PLANES

Radio-controlled (RC) aircraft are flown using a handheld radio transmitter. A user on the ground acts as the aerobatic pilot, moving a joystick to signal maneuvers to the aircraft. Expert RC pilots send their crafts from steep climbs into loops, rolls, and spins.

An RC plane performs a corkscrew maneuver.

RC aircraft are copies of their real, full-size counterparts at 20-50 percent of the size.

CHAPTER 10

BECOMING AN AEROBATIC PILOT

A pilot's license and an aerobatic rating are needed to become an aerobatic performer. Aviators must be able to withstand great **g-forces**. They need to be willing to take risks, but also to be safe. Most aerobatic pilots have been flying for many years. They know their aircraft and what to do if something goes wrong.

For safety, an aerobatic pilot wears a harness, helmet, goggles, and carries a parachute on board.

CHAPTER 11

THE FUTURE

Engineers continue to produce faster, safer, and nimbler aircraft. Aerobatic pilots are excited to push their planes to the limit. New maneuvers are always being developed with stunts that challenge aviators and wow crowds on the ground.

The US Navy Blue Angels began flying new F/A-18 Super Hornets in 2021.

XTREME CHALLENGE

TAKE THE QUIZ BELOW AND PUT WHAT YOU'VE LEARNED TO THE TEST!

1) What two words are combined to make the term aerobatic?

2) Who was one of the first pilots to fly a plane for entertainment purposes? In what year and by what two famous New York landmarks did he fly?

3) What is an ace?

4) What is inverted flying?

5) What are the two categories of aerobatic aircraft? How do they differ?

6) What company produces single- and double-seater aerobatic biplanes?

7) What are the names of two US military formation flying teams? What types of jets do they fly?

GLOSSARY

ace – A highly-skilled pilot. The term first appeared in French newspapers in 1915 to describe World War I aviator Adolphe Pegoud after he became the first pilot to shoot down five enemy aircraft.

barnstormer – A pilot who travels around the country giving flying shows and performing aeronautical stunts, often in rural areas. Farmers' fields became runways, and barns were a place to gather, thus "barnstormer."

biplane – A plane that has two sets of wings, one on top of the other.

dogfight – Fighter planes that duel over battlefields. They make aerobatic turns and shoot at each other.

fragile – Easily broken.

g-force – A unit of force placed on a body when it is subjected to acceleration. The force is felt as increasing weight. The pull of Earth's gravity equals one g-force.

reconnaissance – To explore or scout an enemy's position.

rotor – The spinning blades that provide lift for a helicopter.

sonic boom – A loud noise like an explosion that occurs when a number of sound waves are forced together, such as when a plane travels faster than the speed of sound.

sound barrier – When a plane goes faster than the speed of sound (Mach 1), it is said to have broken the sound barrier.

variant – A different model of the same type of plane, car, or other machine.

World War I – A war that was fought in Europe from 1914 to 1918, involving countries around the world. The United States entered the war in April 1917.

ONLINE RESOURCES

Booklinks
NONFICTION NETWORK
FREE! ONLINE NONFICTION RESOURCES

To learn more about aerobatic aircraft, please visit **abdobooklinks.com** or scan this QR code. These links are routinely monitored and updated to provide the most current information available.

INDEX

A
Aerobat, 16
Aerobatic Team, US, 22
Air Force, US, 34
Aviat Aircraft, 28

B
Blue Angels, 36, 44
Bonhomme, Paul, 24
Bo-105, 38

C
Cessna, 16
Challenger Class, 20
Claw, 28
Curtiss, 10

E
Edge 540
Extra, 18, 19

F
F/A-18 Super Hornets, 36, 44
Fitzgerald, Aaron, 39
F-16 Fighting Falcons, 34

G
Game Composites, 33
GB1 GameBird, 32, 33
Germany, 18
Grant's Tomb, 7

J
Jenny, 10
JN-4, 10
J-1, 10
Jones, Steve, 24

M
MBB, 38

N
Navy, US, 36, 44
New York, 6, 7, 39
New York City, NY, 39

P
Pitts, 22, 23, 28, 29, 29
Pitts, Curtis, 22

R
Red Bull, 39
Red Bull Air Race World Championship, 20
Red Bull Matador Aerobatic Team, 24
Red Star plane, 31
Roy, Gladys, 11
Russia, 26, 30

S
Sbach 300, 25
South African National Aerobatic Championships, 33
Special S1, 22
Standard, 10
Statue of Liberty, 7, 38
Stinkers, 23
S-2C, 28, 29
Suderman, Spencer, 22
Sukhoi, 17, 26
Sunbird S1X, 22
Su-29, 26, 27
Su-26M, 17

T
330SC, 18, 19
Thunderbirds, 34
Timm, Wally, 10

U
Unger, Ivan, 11
United States, 31

W
World Championship, 22
World War I, 9, 10
Wright, Orville, 6
Wright, Wilbur, 6, 7

X
XA41, 24, 25
XtremeAir, 25

Y
Yak-52, 30, 31
Yakovlev, Aleksandir, 30

Z
Zivko Aeronautics, 20